PHILOSOPHY & FUN OF ALGEBRA

BY

MARY EVEREST BOOLE

AUTHOR OF
"PREPARATION OF THE CHILD FOR SCIENCE," ETC.

LONDON: C. W. DANIEL, LTD.
3 Tudor Street, E.C. 4.

ISBN   1-60386-126-2

𝕿𝔬

BASIL AND MARGARET

—————

MY DEAR CHILDREN,

A young monkey named Genius picked a green walnut, and bit, through a bitter rind, down into a hard shell. He then threw the walnut away, saying: "How stupid people are! They told me walnuts are good to eat."

His grandmother, whose name was Wisdom, picked up the walnut—peeled off the rind with her fingers, cracked the shell, and shared the kernel with her grandson, saying: "Those get on best in life who do not trust to first impressions."

In some old books the story is told differently; the grandmother is called Mrs Cunning-Greed, and she eats all the kernel herself. Fables about the Cunning-Greed family are written to make children laugh. It is good for you to laugh; it makes you grow strong, and gives you the habit of understanding jokes and not being made miserable by them. But take care not to believe such fables; because, if you believe them, they give you bad dreams.

MARY EVEREST BOOLE.

January 1909.

Contents

Chapter 1

From Arithmetic To Algebra

Arithmetic means dealing logically with facts which we know (about questions of number).

"Logically"; that is to say, in accordance with the "Logos" or hidden wisdom, *i.e.* the laws of normal action of the human mind.

For instance, you are asked what will have to be paid for six pounds of sugar at 3d. a pound. You multiply the six by the three. That is not because of any property of sugar, or of the copper of which the pennies are made. You would have done the same if the thing bought had been starch or apples. You would have done just the same if the material had been tea at 3s. a pound. Moreover, you would have done just the same *kind* of action if you had been asked the price of seven pounds of tea at 2s. a pound. You do what you do under direction of the Logos or hidden wisdom. And this law of the Logos is made not by any King or Parliament, but by whoever or whatever created the human mind. Suppose that any Parliament passed an act that all the children in the kingdom were to divide the price by the number of pounds; the Parliament could not make the answer come right. The only result of a foolish Act of Parliament like that would be that everybody's sums would come wrong, and everybody's accounts be in confusion, and everybody would wonder why the trade of the country was going to the bad.

In former times there were kings and emperors quite stupid enough to pass Acts like that, but governments have grown wiser by experience and found out that, as far as arithmetic goes, there is no use in ordering people to go contrary to the laws of the Logos, because the Logos has the whip hand, and knows its own business, and is master of the situation. Therefore children now are allowed to study the laws of the Logos, whenever the business on hand is finding out how much they are to pay in a shop.

Sometimes your teachers set you more complicated problems than:—What is the price of six pounds of sugar? For instance:—In what proportion must one mix tea bought at 1s. 4d. a pound with tea bought at 1s. 10d. a pound so as to make 5 per cent. profit by selling the mixture at 1s. 9d. a pound?

Arithmetic, then, means dealing logically with certain facts that we know, about number, with a view to arriving at knowledge which as yet we do not possess.

When people had only arithmetic and not algebra, they found out a surprising amount of things about numbers and quantities. But there remained problems which they very much needed to solve and could not. They had to guess the answer; and, of course, they usually guessed wrong. And I am inclined to think they disagreed. Each person, of course, thought his own guess was nearest to the truth. Probably they quarrelled, and got nervous and overstrained and miserable, and said things which hurt the feelings of their friends, and which they saw afterwards they had better not have said—things which threw no light on the problem, and only upset everybody's mind more than ever. I was not there, so I cannot tell you exactly what happened; but quarrelling and disagreeing and nerve-strain always do go on in such cases.

At last (at least I should suppose this is what happened) some man, or perhaps some woman, suddenly said: "How stupid we've all been! We have been dealing logically with all the facts we knew about this problem, except the most important fact of all, the fact of our own ignorance. Let us include that among the facts we have to be logical about, and see where we get to then. In this problem, besides the numbers which we do know, there is one which we do not know, and which we want to know. Instead of guessing whether we are to call it nine, or seven, or a hundred and twenty, or a thousand and fifty, let us agree to call it x, and let us always remember that x stands for the Unknown. Let us write x in among all our other numbers, and deal logically with it according to exactly the same laws as we deal with six, or nine, or a hundred, or a thousand."

As soon as this method was adopted, many difficulties which had been puzzling everybody fell to pieces like a Rupert's drop when you nip its tail, or disappeared like bats when the sun rises. Nobody knew where they had gone to, and I should think that nobody cared. The main fact was that they were no longer there to puzzle people.

A little girl was once saying the Evening Hymn to me, "May no ill dreams disturb my rest, No powers of darkness me molest." I asked if she knew what *Powers of Darkness* meant. She said, "The wolves which I cannot help fancying are under my bed when all the time I know they are not there. They must be the Powers of Darkness, because they go away when the light comes."

Now that is exactly what happened when people left off disputing about what they did not know, and began to deal logically with the fact of their own ignorance. This method of solving problems by honest confession of one's ignorance is called Algebra.[1]

The name Algebra is made up of two Arabic words.

The science of Algebra came into Europe through Arabs, and therefore is

---

[1] *See* Appendix.

called by its Arabic name. But it is believed to have been known in India before the Arabs got hold of it.

Any fact which we know or have been told about our problem is called a datum. The number of pounds of sugar we are to buy is one datum; the price per pound is another.

The plural of datum is data. It is a good plan to write all one's data on one column or page of the paper and work one's sum on the other. This leaves the first column clear for adding to one's data if one finds out any fresh one.

Chapter 2

The Making of Algebras

The Arabs had some cousins who lived not far off from Arabia and who called themselves Hebrews. A taste for Algebra seems to have run in the family. Three Algebras grew up among the Hebrews; I should think they are the grandest and most useful that ever were heard of or dreamed of on earth.

One of them has been worked into the roots of all our science; the second is much discussed among persons who have leisure to be very learned. The third has hardly yet begun to be used or understood in Europe; learned men are only just beginning to think about what it really means. All children ought to know about at least the first of these.

But, before we begin to talk about the Hebrew Algebras, there are two or three things that we must be quite clear about.

Many people think that it is impossible to make Algebra about anything except number. This is a complete mistake. We make an Algebra whenever we arrange facts that we know round a centre which is a statement of what it is that we want to know and do not know; and then proceed to deal logically with all the statements, including the statement of our own ignorance.

Algebra can be made about anything which any human being wants to know about. Everybody ought to be able to make Algebras; and the sooner we begin the better. It is best to begin before we can talk; because, until we can talk, no one can get us into illogical habits; and it is advisable that good logic should get the start of bad.

If you have a baby brother, it would be a nice amusement for you to teach him to make Algebra when he is about ten months or a year old. And now I will tell you how to do it.

Sometimes a baby, when it sees a bright metal tea-pot, laughs and crows and wants to play with the baby reflected in the metal. It has learned, by what is called "empirical experience," that tea-pots are nice cool things to handle. Another baby, when it sees a bright tea-pot, turns its head away and screams, and will not be pacified while the tea-pot is near. It has learned, by empirical experience, that tea-pots are nasty boiling hot things which burn one's fingers.

Now you will observe that both these babies have learnt by experience.

Some people say that experience is the mother of Wisdom; but you see that both babies cannot be right; and, as a matter of fact, both are wrong. If they could talk, they might argue and quarrel for years; and vote; and write in the newspapers; and waste their own time and other people's money; each trying to prove he was right. But there is no wisdom to be got in that way. What a wise baby knows is that he *cannot tell*, by the mere look of a tea-pot, whether it is hot or cold. The fact that is most prominent in his mind when he sees a tea-pot is the fact that *he does not know* whether it is hot or cold. He puts that fact along with the other fact:—that he would very much like to play with the picture in the tea-pot supposing it would not burn his fingers; and he deals logically with both these facts; and comes to the wise conclusion that it would be best to go very cautiously and find out whether the tea-pot is hot, by putting his fingers near, but not too near. That baby has begun his mathematical studies; and begun them at the right end. He has made an Algebra for himself. And the best wish one can make for his future is that he will go on doing the same for the rest of his life.

Perhaps the best way of teaching a baby Algebra would be to get him thoroughly accustomed to playing with a bright vessel of some kind when cold; then put it and another just like it on the table in front of him, one being filled with hot water. Let him play with the cold one; and show him that you do not wish him to play with the other. When he persists, as he probably will, let him find out for himself that the two things which look so alike have not exactly the same properties. Of course, you must take care that he does not hurt himself seriously.

Chapter 3

Simultaneous Problems

It often happens that two or three problems are so entangled up together that it seems impossible to solve any one of them until the others have been solved. For instance, we might get out three answers of this kind:—

x equals half of y;
y equals twice x;
z equals x multiplied by y.

The value of each depends on the value of the others.

When we get into a predicament of this kind, three courses are open to us.

We can begin to make slap-dash guesses, and each argue to prove that his guess is the right one; and go on quarrelling; and so on; as I described people doing about arithmetic before Algebra was invented.

Or we might write down something of this kind:—

The values cannot be known. There is no answer to our problem.

We might write:—

x is the unknowable;
y is non-existent;
z is imaginary,

and accept those as answers and give them forth to the world with all the authority which is given by big print, wide margins, a handsome binding, and a publisher in a large way of business; and so make a great many foolish people believe we are very wise.

Some people call this way of settling things Philosophy; others call it arrogant conceit. Whatever it is, it is not Algebra. The Algebra way of managing is this:—

We say: Suppose that x were Unity (1); what would become of y and z? Then we write out our problem as before; only that, wherever there was x, we now write 1.

If the result of doing so is to bring out some such ridiculous answer as "2 and 3 make 7," we then know that x cannot be 1. We now add to our column of data, "x cannot be 1."

But if we come to a truism, such as "2 and 3 make 5," we add to our column of data, "x may be 1." Some people add to their column of data, "x is 1," but that again is not Algebra. Next we try the experiment of supposing x to be equal to zero (0), and go over the ground again.

Then we go over the same ground, trying y as 1 and as 0.

And then we try the same with z. Some people think that it is waste of time to go over all this ground so carefully, when all you get by it is either nonsense, such as "2 and 3 are 7"; or truisms, such as "2 and 3 are 5." But it is not waste of time. For, even if we never arrive at finding out the value of x, or y, or z, every conscientious attempt such as I have described adds to our knowledge of the structure of Algebra, and assists us in solving other problems.

Such suggestions as "suppose x were Unity" are called "working hypotheses," or "hypothetical data." In Algebra we are very careful to distinguish clearly between actual data and hypothetical data.

This is only part of the essence of Algebra, which, as I told you, consists in preserving a constant, reverent, and conscientious awareness of our own ignorance.

When we have exhausted all the possible hypotheses connected with Unity and Zero, we next begin to experiment with other values of x; *e.g.*—suppose x were 2, suppose x were 3, suppose it were 4. Then, suppose it were one half, or one and a half, and so on, registering among our data, each time, either "x may be so and so," or "x cannot be so and so."

The method of finding out what x cannot be, by showing that certain suppositions or hypotheses lead to a ridiculous statement, is called the method of *reductio ad absurdum.* It is largely used by Euclid.

Chapter 4

Partial Solutions and the Provisional Elimination of Elements of Complexity

Suppose that we never find out for certain whether x is unity or zero or something else, we then begin to experiment in a different direction. We try to find out which of the hypothetical values of x throw most light on other questions, and if we find that some particular value of x—for instance, unity—makes it easier than does any other value to understand things about y and z, we have to be very careful not to slip into asserting that x *is* unity. But the teacher would be quite right in saying to the class, "For the present we will leave alone thinking about what would happen if x were something different from unity, and attend only to such questions as can be solved on the supposition that x is unity." This is what is called in Algebra "provisional elimination of some elements of complexity."

It might happen that one of the older pupils, specially clever at mathematics, but not very well disciplined, should start some point connected with the supposition that x is something different than unity. It would be the teacher's business to remind her: "At present we are dealing with the supposition that x *is* unity. When we have exhausted that subject we will investigate your question. But, till then, please do not distract the attention of the class by talking about what is not the business on hand at present."

If the girl forgot, the teacher might say: "I should very much like you to try your own suggestion in private, but please do not talk about it in class till I give you leave."

If she forgot again, the teacher might say,—I think I should be inclined to say:—"If you cannot remember not to distract the class by talking about what is irrelevant to the business on hand, I shall have to request you to keep outside my class-room till you can."

In an orderly school the teachers have time to be polite, and it is their business to set the example of being so. In history, especially such history as that of half-civilised countries 3000 years ago, teachers were under too much strain to cultivate either a polite *manner* of saying things, or, what is of far more consequence, that genuine intellectual courtesy which is the absolutely necessary condition for the development of any really perfect mathematical system. The great Hebrew Algebra, therefore, never became quite perfect. It was only rough hewn, so to speak; and its manners and customs were rough too. The teachers had ways of saying, "Hold your tongue, or else go out of my class-room," which perhaps we should now call bigoted and brutal. But what I want you to notice is that "Hold your tongue, or get out of my class-room," is not the same thing as "My hypothesis is right, and yours ought not to be tried anywhere."

This latter is contrary to the essential basis of Algebra, viz., a recognition of one's own ignorance.

The other, a rough way of saying "Get out of my class-room," is only contrary to that fine intellectual courtesy which is essential to the *perfection* of mathematical method.

Chapter 5

Mathematical Certainty
and Reductio ad Absurdum

It is very often said that we cannot have mathematical certainty about anything except a few special subjects, such as number, or quantity, or dimensions.

Mathematical certainty depends, not on the subject matter of our investigation, but upon three conditions. The first is a constant recognition of the limits of our own knowledge and the fact of our own ignorance. The second is reverence for the As-Yet-Unknown. The third is absolute fearlessness in meeting the *reductio ad absurdum*. In mathematics we are always delighted when we come to any such conclusion as $2 + 3 = 7$. We feel that we have absolutely cleared out of the way one among the several possible hypotheses, and are ready to try another.

We may be still groping in the dark, but we know that one stumbling-block has been cleared out of our path, and that we are one step "forrader" on the right road. We wish to arrive at truth about the state of our balance sheet, the number of acres in our farm, the time it will take us to get from London to Liverpool, the height of Snowdon, the distance of the moon, and the weight of the sun. We have no desire to deceive ourselves upon any of these points, and therefore we have no superstitious shrinking from the rigid *reductio ad absurdum*. On some other subjects people do wish to be deceived. They dislike the operation of correcting the hypothetical data which they have taken as basis. Therefore, when they begin to see looming ahead some such ridiculous result as $2 + 3 = 7$, they shrink into themselves and try to find some process of twisting the logic, and tinkering the equation, which will make the answer come out a truism instead of an absurdity; and then they say, "Our hypothetical premiss is most likely true because the conclusion to which it brings us is obviously and indisputably true."

If anyone points out that there seems to be a flaw in the argument, they say, "You cannot expect to get mathematical certainty in this world," or "You must not push logic too far," or "Everything is more or less compromise," and so on.

Of course, there is no mathematical certainty to be had on those terms. You could have no mathematical certainty about the amount you owed your grocer if you tinkered the process of adding up his bill. I wish to call your attention to the fact that *even in this world* there is a good deal of mathematical certainty to be had by whosoever has endless patience, scrupulous accuracy in stating his own ignorance, reverence for the As-Yet-Unknown, and perfect fearlessness in meeting the *reductio ad absurdum*.

Chapter 6

The First Hebrew Algebra

The first Hebrew algebra is called Mosaism, from the name of Moses the Liberator, who was its great Incarnation, or Singular Solution. It ought hardly to be called an algebra: it is the master-key of all algebras, the great central director for all who wish to learn how to get into right relations to the unknown, so that they can make algebras for themselves. Its great keynotes are these:—

When you do not know something, and wish to know it, state that you do not know it, and keep that fact well in front of you.

When you make a provisional hypothesis, state that it is so, and keep that fact well in front of you.

While you are trying out that provisional hypothesis, do not allow yourself to think, or other people to talk to you, about any other hypothesis.

Always remember that the use of algebra is to *free people from bondage.* For instance, in the case of number: Children do their numeration, their "carrying," in tens, because primitive man had nothing to do sums with but his ten fingers.

Many children grow superstitious, and think that you cannot carry except in tens; or that it is wrong to carry in anything but tens. The use of algebra is to free them from bondage to all this superstitious nonsense, and help them to see that the numbers would come just as right if we carried in eights or twelves or twenties. It is a little difficult to do this at first, because we are not accustomed to it; but algebra helps to get over our stiffness and set habits and to do numeration on any basis that suits the matter we are dealing with.

Of course, we have to be careful not to mix two numerations. If we are working a sum in tens, we must go on working in tens to the end of that sum.

Never let yourself get fixed ideas that numbers (or anything else that you are working at) will not come right unless your sum is set or shaped in a particular way. Have a way in which you usually do a particular kind of sum, but do not let it haunt you.

You may some day become a teacher. If ever you are teaching a class how to set down a sum or an equation, say "This is my way," or "This is the way which I think you will find most convenient," or "This is the way in which the Government Inspector requires you to do the sums at present, and therefore you must learn it." But do not take in vain the names of great unseen powers to back up either your own limitations, or your own authority, or the Inspector's authority. Never say, or imply, "Arithmetic requires you to do this; your sum will come wrong if you do it differently." Remember that arithmetic requires nothing from you except absolute honesty and patient work. You get no blessing from the Unseen Powers of Number by slipshod statements used to make your own path easy.

Be very accurate and plodding during your hours of work, but take care not to go on too long at a time doing mere drudgery. At certain times give yourself a full stretch of body and mind by going to the boundless fairyland of your subject. Think how the great mathematicians can weigh the earth and measure the stars, and reveal the laws of the universe; and tell yourself that it is all one science, and that you are one of the servants of it, quite as much as ever Pythagoras or Newton were.

Never be satisfied with being up-to-date. Think, in your slack time, of how people before you did things. While you are at school my little book, *Logic of Arithmetic*, will help you to find out many things about your ancestors which may amuse and interest you; but, as soon as you leave school and choose your own reading, take care to read up the histories of the struggles and difficulties of the people who formerly dealt with your own subject (whatever that may be).

If you find the whole of the data too complicated to deal with, and judge that it is necessary to eliminate one or more of them, in order to reduce your material within the compass of your own power to manage, do it as a *provisional* necessity. Take care to register the fact that you have done so, and to arrange your mind, from the first, on the understanding that the eliminated data will have to come back. Forget them during the working out of your experimental equation; but never give way to the feeling that they are got rid of and done with.

Be very careful not to disturb other people's relationships to each other. For instance, if a teacher is explaining something to another pupil, never speak till she has done. Beware of the sentimental craving to be "in it." Any studying-group profits by right working relations being set up between any two members; and ultimately each member profits. The whole group suffers from any distraction between any two. Therefore listen and learn what you can; but never disturb or distract.[1]

Take care not to become a parasite; do not lazily appropriate the results of other people's labour, but learn and labour truly to get your own living. Take care that everything you possess, whether physical, mental, or spiritual, shall be the result of your own toil as well as other people's; and remember that you are bound to pay, in some shape or way, everyone who helps you.

Do not make things easy for yourself by speaking or thinking of data as if they were different from what they are; and do not go off from facing data as they are, to amuse your imagination by wishing they were different from what

---

[1] D. Marks bases the Seventh Commandment on the desirability of not distracting existing relations.

they are. Such wishing is pure waste of nerve force, weakens your intellectual power, and gets you into habits of mental confusion.

When the time comes to stop grind-work, there is no better rest than amusing your imagination by thinking of non-existent possibilities; but do it on a free, generous scale. Give yourself a perfectly free rein in the company of the Infinite. During such exercise of the imagination, remember that you are in the company of the Infinite, and are not dealing with, or tinkering at, the problem on your paper.

Keep always at hand, clearly written out, a good standard selection of the most important formulæ—Arithmetical, Algebraic, Geometric, and Trigonometrical, and accustom yourself to test your results by referring to it.

These are the main laws of mathematical self-guidance. Once upon a time "Moses" projected them on to the magic-lantern screen of legislation. In that form they are known as the Ten Commandments; or, to change the metaphors, we might call the Ten Commandments the outer skin of the mathematical body.

A great many people seem to suppose that, though everyone ought to keep the Ten Commandments, it does not matter what happens to one's mind. Just so, there are people who live unhealthy lives, and think they can make all right by putting cosmetics on their skin. But I hope you have learned in the hygiene class how stupid and futile all that is. The way to have a healthy skin is to grow it, by leading a hygienic life on a moderate allowance of pure wholesome food, and taking a proper amount of exercise in pure fresh air. People who do that with their minds grow the Ten Commandments naturally, just as Moses grew them. The world has been trying the other plan—bad food and air inside, and cosmetics outside—for at least 4000 years; and not much seems to have come of it yet. The Ten Commandments have not yet succeeded in getting themselves kept. Perhaps that is why some schoolmasters and mistresses think they would like to try the other plan now. Still, it is very good to have a normal model of what a healthy human being ought to look like outside. It is good to have a standard for reference. Therefore do not get too much immersed in the mere details of your own problems. Learn the Ten Commandments and a few other old standard formularies by heart, and repeat them every now and then. And say to yourself, "If I really am doing my algebra quite rightly, *this* (the standard formularies) is how I shall think and feel and wish. I shall wish to behave thus, not because anybody ordered me to do so, but from sheer liking and sense of the general fitness of things."

Chapter 7

How to Choose Our Hypotheses

The faculties by means of which we get our positive data are called the senses (sight, hearing, etc.).

The faculty by means of which we get our hypothetical data is called the Imagination.

Some persons are prone to warn young people against what they call an excessive exercise of the imagination. Of course, to say that "excessive" anything is too much is a mere truism, but nobody knows yet what is the proper amount of use for the imagination. What we do know is that there is a good deal of excessive mis-use of the imagination, by which I mean that there is a frightful amount of using it contrary to the laws of its normal action. A kind of use of it, such as, when we find a child doing it with its eyes, we say, "Do not learn the habit of squinting"; or if it does the analogous thing with its legs, we say, "Go and run about, or do some gymnastics; do not stand there lolloping crooked against the wall."

Squinting and lolloping crooked are things that it is best to avoid doing much of with any part of one's self.

Moreover, it is bad to spend too many hours over either a microscope or a telescope, or in gazing fixedly at some one-distance range. The eyes need change of focus. So does the imagination.

There has been in modern Europe a shocking riot in mis-use of the imagination. The remedy is to learn to use it. But the same kind of people who would like to bandage a child's eyes lest it should learn to squint, like to bandage the imagination lest it should wear itself out by squinting.

In a school which professes to be conducted on hygienic principles, we have nothing to do with that sort of pessimistic quackery. We use the imagination as freely as the hands and eyes.

But when we come to the end of our arithmetic we do not content ourselves with guesses; we proceed to algebra–that is to say, to dealing logically with the fact of our own ignorance. One of the data that we do know is that all great nerve-centres affect each other. Mis-use of any one tends more or less to produce distorted action in the others. And, quite apart from that consideration, any energetic and continued action of one tends more or less to suppress the action of the others, for the time being, by drawing the blood from the organs which are the seat of them; and then, when normal circulation is restored, to produce for a time an unusual sensitiveness in the others. There is nothing abnormal or wrong in this, provided that we recognise the fact, and, as I said, are careful to deal logically with the fact of our own ignorance whenever anything happens either to our eyes or to our imagination which we do not at the moment quite understand.

If you ever arrive at using your imagination strongly and rightly in the construction of any sort of algebra, you may find that it affects to some extent your sense-organs. It certainly will affect them more or less whether you know it or not. What I mean is that it may affect them in a way that forces you to be aware of the fact. If ever this should happen, take it quite naturally; and as long as you are too young to understand how it happens, just say to yourself, "This is x, one of the things that I do not know, and perhaps shall know some day if I go on quietly acting in accordance with strict logic, and remembering my own ignorance."

The ancient Hebrews used their imaginations very freely, and sometimes really very logically. And sometimes the free use of the imagination produced sensations in the eyes and ears as if of seeing and hearing. They considered this quite natural, as it really was. Many great mathematicians in modern Europe have had these sensations.

The Hebrews called these sensations by a Hebrew word which is translated by the English word "angel," from the Greek "angelos," a messenger. The Hebrews were quite right. The sensations are messengers from the Great Unknown. They bring no information about outside facts. No angel tells you how many petals there are in a buttercup; if you want to know that, you are supposed to ask the buttercup itself. No angel tells you the price of sugar; you ought to ask your grocer. No angel tells you how to invest your money; you ought to ask your banker or your lawyer. There are people foolish enough to ask angels about investments, or about which horse will win a race; which is just as foolish as asking your banker in town how many blossoms there are on the rose tree in your country garden. It is not his business, and if he made a guess it would most likely turn out a wrong one. All that sort of thing is quackery and superstition.

But the angels do bring us very reliable information from a vast region of valuable truth about which most of us know very little as yet. They guide us how to frame our *next provisional working hypothesis*, how to choose the particular hypothesis which at our present stage of knowledge and development will be most illuminating for us. Some of the angels come during sleep; we call them dreams. Dreams sometimes suggest the best working hypothesis to experiment on next. More often they warn us against thinking upon some hypothetical basis which for the present will not suit us.

And here comes in the value of such formulæ as the Ten Commandments.

They are the laws of the *normal* working of the brain machinery.

The angel (or imaginary messenger) suggests to you the one among possible working hypotheses on which your brain will most readily work. Now the formularies of which I spoke give you the laws of healthy brain action. Therefore, if the angel suggests something contrary to the registered formulas, he is suggesting the hypothesis which you ought carefully to avoid thinking out or using at that time. It is of all paths towards disease the one which will lead you, in your present condition, most rapidly towards disease. But if the imaginary angel suggests nothing contrary to the formularies, then the image or idea which he suggests is likely to be one on which your mind for the time being can work safely, and *the* one along which it can work most easily and profitably.

When your imagination is acting strongly in providing you with working hypotheses, there are a few little precautions which you ought to observe.

Do not at such times take either very rapid or very much prolonged physical exercise.

Be rather particular not to eat anything either indigestible or highly flavoured.

Even if you were in the habit of taking any kind of alcoholic stimulant (which, while you are young, I hope you will not do), avoid it during the process of framing hypotheses. Be extra careful, at such times, to keep up any routine exercises of slack muscles and slow breathing which you find suit you.

Take a little extra care, at such times, not to catch cold. You are rather less liable than usual to take cold at such times; but, on the other hand, you are less conscious than usual of ordinary physical sensations, and may be very cold without knowing it. A chill may settle locally, and produce permanent mischief.

Above all, be very careful, while the imaginative fit is on, to avoid letting the subject as to which your imagination is stirred become the object of either fun, vanity, or gossip. The vision which you see may quite harmlessly and legitimately become a source of fun to yourself and your friends at some future time, but take care never to gossip or joke about it until it has passed from the condition of imaginative vision to that of working hypothesis. But the most important precaution of all is incessant reverence for the Great Unknown, the sacred x: or, in other words, a constant awareness of your own ignorance.

Remember always that Genius means conscientious, careful work on suggestions of the imagination taken as provisional hypotheses.

To take suggestions of the Imagination as fact is Insanity. When you hear of a man that he has unquestionable genius but is a little mad, that means that he sometimes takes the products of his imagination as working hypotheses, but sometimes mistakes them for facts.

All the above precautions may be summed up in one sentence: Remember that the more active the imagination is, the less the physical and moral instincts are on the alert; therefore, conscious precaution should supplement instinct at such times, until self-protection has become so fixed by habit as to become in its turn automatic and instinctive.

If you observe these precautions you need not fear using your imagination freely. When you hear of some brilliant imaginative writer who has come to grief physically, mentally, or morally, after a short and brilliant career, you will find it advantageous to try to find out which of the precautions he has been neglecting.

In future letters I hope to point out to your notice some famous cases of disaster due to such neglect.

Chapter 8

The Limits of the Teacher's Function

One of the greatest causes of mental and moral confusion, as well as of absolute insanity, in modern Europe, is the fact that numbers of people plunge into the second and third great Hebrew algebras before they rightly understand the first. Even if they are silent about their results, this distracts their own minds, and sows the seeds of bad habits and mental confusion in their own constitutions. Many of these people give to the world their own wild guesses about the second and third algebras, and that puts the rest of the world into confusion. We are, therefore, not going to enter on the question of the second algebra till I have provided you with the possibility of understanding and practising the first. In the next few chapters I hope to give you a series of stories of people who used, and sometimes mis-used, the algebra of Moses, in order that you may see how to work the rules strictly and how mistakes might creep in.

But, before we begin our stories, there is one principle to which I must call your attention: it is the business of your teachers at school to see that you acquire skill in using certain implements or tools; it is not their business nor mine to decide what use you shall make, when you are grown up, of the skill which you have acquired. It is their business to see that you learn to read and to speak properly; it is not their business to decide beforehand whether you shall recite in public or only read to your own family and your sick friends. It is their business to see that you know how to sew; but not to settle whether you shall, in future, make your own clothes or work for the poor. So it is with the tools of the mind, such as algebra and logic. It is our business to see that you know how to use algebraic and logical method accurately and skilfully; it is not our business to decide whether, in the future, you shall use your skill to deceive other people or to show them the truth. It *is* our business to see that you do not deceive yourself, because deceiving *yourself* distorts your brain and ruins the possibility of using logical methods skilfully to arrive at the knowledge of truths.

When you have found out a truth, then the question whether you shall or shall not tell it to other people is a matter of conscience. You will have to settle it alone with the Great Power which no man knows. Self-deception, slipshod logic, and bad algebra are things which it is the business of your elders to protect you from while you are young, in order that you may not *lose the power* of being honest in case you wish to be so. My business is not to judge what is good or bad conduct, but to see that you learn how to be perfectly honest with yourself. I wish you to notice this, because in the books of the Hebrew algebra you will sometimes find good kind people spoken of very harshly; and some of the most dishonest and selfish people in the world praised and spoken of as blessed. This puzzles many good people, because they choose to fancy that the Hebrew books are sermons about right and wrong feelings; and do not like to recognise that they are really about the algebra of logic.

As I said before, people who really conduct their minds strictly according to the algebra of logic are very prone to grow kindness and honesty towards other people, without thinking about it, as a matter of taste, of choice. They *like* being kind and honest better than being selfish and dishonest, and they become kind and honest without thinking much about it. But honesty to other people and honesty to yourself *are* two different things, and must be kept apart in your mind, just as, in physiology class, you keep apart the flesh of an animal and its skin. You believe that if the flesh is thoroughly healthy it will grow a good skin; but, while you are studying, you do not mix up statements about the one with guesses about the other. If we find that a man's logic was good, and his conduct what we should call bad, we must do what a doctor would do if he found a spot on a patient's skin which he could not account for by anything wrong in his circulation or digestion. He ought not to say either, "That spot is not there," or, "I suppose it is right that spot should be there," nor, on the other hand, to jump to the conclusion that that patient had been eating some particularly unwholesome thing. He ought to register in his mind, as one of his data, the fact of his own ignorance of how that spot came there. I shall have to tell you in another chapter the story of one of the most selfish and deceitful persons that ever lived, as to his conduct towards other people, but who was said to be blessed, apparently for no reason except that he was absolutely straight with his logic and honest with himself.

Besides, no one who is consciously and deliberately dishonest to serve his own selfish purposes can ever do as much harm to other people as is done every day by men and women who have muddled their own brains with crooked logic.

Chapter 9

The Use of Sewing Cards

When you go for holidays perhaps your friends will ask you what is the use of sewing curves on cards. I should like you to know exactly what to say.

The use of the single sewing cards is to provide children in the kindergarten with the means of finding out the exact nature of the relation between one dimension and two.

There is another set of sewing cards which is made by laying two cards side by side on the table and pasting a tape over the crack between them. This tape forms a hinge. You can lay one card flat and stand the other edgeways upright, and lace patterns between them from one to the other.

The use of this part of the method is to provide girls in the higher forms with a means of learning the relation between two dimensions and three.

There is another set of models, the use of which is to provide people who have left school with a means of learning the relation between three dimensions and four.

The use of the books which are signed George Boole or Mary Everest Boole is to provide reasonable people, who have learned the logic of algebra conscientiously, with a means of teaching themselves the relations between n dimensions and $n + 1$ dimensions, whatever number n may be.

The above is a quite accurate account of the real Boole Method; as much as there is any need for you to know while you are at school.

I should feel grateful to you if you will each copy it out in a clear handwriting, and keep it by you, and take it home whenever you go away from school for the holidays. It would be all the better if you learned it by heart.

And now I will tell you why I am so anxious about this.

The Boole method is a conveyance which will take you safely to wherever the Great Unknown directs you to go. Some people mistake it for the carpet in the *Arabian Nights*, which took whoever stepped on it wherever he or she *wished* to go–which is a quite different thing. The true Boole method depends essentially on making a right use of imaginary hypotheses. The magic carpet depends for its efficacy on making a wrong use of imaginary hypotheses.

People get to very queer places on that carpet. I have been for several excursions on it, so I know.

One of the places it can take you to is a town where all the front doors open on to a street very like Regent Street; with the most gorgeous millinery, jewellery, and fruits in shop windows; and all the back doors open to wild country where blue roses, black tulips, and the fattest double carnations of all colours (including green ones) grow wild in the hedges and fields; and where all the pigs have wings.

Another place that it can take you to is one where pigs can wallow in all the filth they like without soiling their wings; and moths fly into candles without singeing theirs.

The carpet will take you straight *to* whatever place you wish to go to. It is by no means warranted to take you safely back.

The advantage of Boole's method is that it *is* warranted to bring you safe down somewhere on solid earth,—not always the exact place you started from, but a safe and clean place of some kind—and to deposit you steady on your feet, with a compass in your pocket which will show you a straight way home.

Chapter 10

The Story of a Working Hypothesis

In an old Hebrew book there is a story of a person named Jacob, which means the Supplanter. If you want to know why, you had better read the story for yourself some day. It is not entirely a pretty story, but it is very instructive. Jacob had a dream in which he saw "angels" coming down a ladder. It would be a very profitable exercise of your imagination to ask yourselves why this particular patriarch saw angels on a ladder, whereas so many other Hebrews saw them in clouds, or flying down on wings, or mixed up with flames and other romantic, pretty, moving things.

Jacob had another dream, and saw an angel who wrestled with him, and apparently left him with sciatica for life; which is not surprising, for he had been sleeping out of doors on bare ground, just when *he* had been wrestling with very serious difficulties caused by his own dishonest tricks. At such times, as I told you before, people had better be a little extra careful not to catch cold; because colds caught under such conditions are rather prone to leave unpleasant traces, which last a long time, and sometimes all one's life.

Well, the angel who gave Jacob sciatica gave him something else: a new name. Why did he give him a new name? Taking a new name was an ancient ceremony which meant entering a new service. Sixty years ago servants in Devonshire were called by their employer's name. A gardener would have two names—his own, which he got from his father, and his master's. I have even heard dogs called by their master's names, for instance, Toby Smith, or Ponto Jones.

You will often notice in old books that when people were converted, that is to say, when they either took up a new religion or turned from bad ways to good ones, the people who persuaded them to be converted gave them a new name, very often the teacher's own name. Well, the angel who wrestled with Jacob appears to have converted him. He seems to have persuaded Jacob that there are other ways of getting on in the world and promoting the fortunes of one's children and grandchildren besides cheating everybody, including one's own nearest relations.

Therefore Jacob was not to be called "the Supplanter" any more: his new name was to be Israël. Jacob's descendants are called Hebrews, and also "the people of Israël." Israël was the new name which Jacob got when he turned from cheating to a better way of getting on in life.

What was that better way? That is our x, our first unknown. What does the word Israël mean? That is our y, our second unknown. I may as well tell you at once that, so far as I am concerned, y remains unknown. I want you to take notice that I do not know what the word Israël means. But some twenty years ago my imagination supplied me with a working hypothesis:–Suppose Israël meant rhythm.

Now if I had gone telling people that *Israël means rhythm*, I should have been contradicted and laughed at and told that I had no proof of what I said and was talking of what I knew nothing about; and whoever said so would have been perfectly right. I should have been cheating myself and getting into bad slipshod habits. What I did was to post up inside my brain as a working hypothesis: "*Suppose* Israël means rhythm, what would be the consequence of that hypothesis?" Then I read through old books of the Hebrews, putting in my mind the word "rhythm" wherever I found the word "Israël," and "the people of rhythm" instead of "the people of Israël."

In the stories that are told about Jacob and his grandfather Abraham the angels are represented as telling the two men that if they would obey the angels, not only they themselves would be blessed, but all their descendants would be blessed too, and be made, at last, the means of conferring a great blessing on all the world; Moses warned them that, if they did not obey their own special angels, some special trouble would come to them.

My imagination suggested to me that perhaps getting into the swing of rhythmic beats is good for all people, but more good for the people of Israël than for anybody else; and that wandering off into irregular un-rhythmic freaks is more bad for the people of Israël than for anybody else.

This, again, you will observe, is purely imaginary hypothesis. I had not the faintest warrant for saying anything of the kind; therefore I did not say it; but I experimented at treating my Hebrew friends and acquaintance *as if* they were natural born ministers, or servants, of the principle of rhythmic beat; as if it was their business to introduce respect for rhythm and an orderly arrangement of time into the general morals of the world; and as if they would, of course, become degraded more than other people, if they allowed themselves to drift into being irregular and disorderly. Now you will observe that, though all this was purely imaginary hypothesis, it was of a harmless kind; there is nothing contrary to the ten commandments, or to any other register of safe rules, in treating one's Hebrew acquaintance as if one expected them to be more orderly as to time than other people.

The registered rules allowed me to consider this a safe road; and my imagination showed me that it was one along which I could travel quickly; therefore I started to go along it and waited to see where I got to. One consequence which came was that some of the people of Israël began telling me that I seemed to know things about their old books (even some old books that I had never read), which they themselves had never observed before; I had enabled them to get at real values for the x's and y's; of some of their problems.

Please notice that all this is pure imaginary hypothesis. Ancient peoples made a hypothesis, for which they had no authority, about angels; and I made one, for which I had no authority, about some of those supposed angels. And, by dealing logically with these imaginations, we got to some very real knowledge.

Chapter 11

Macbeth's Mistake

The whole question of choosing one's next working hypothesis has been fogged, owing to people's neglect of a very simple principle. Suppose you are out bicycling in a strange place. You come to a bit of smooth, good road, which is either flat or goes very gently down hill; and presently curves in a nice, big, easy sweep round a bit of wood or a cliff, so that you cannot see far along it. What you know at once is that you can, *if you choose*, get up great speed without overmuch exertion. That is obvious, and needs no discussion. The question you have to settle is: Shall you choose to do it?

If you have heard the whole road spoken of, in general terms, as a nice safe one to go on, you probably do choose to make use of the specially easy bit of the road to get up a lively spin.

But supposing that, at the beginning of the gentle slope down, you come upon a notice board with an inscription "Go slowly," or "Dangerous to cyclists," I hope you would have sense enough not to think—"What do those old fogies know about the needs of the young generation? I have a right to go fast if I choose, and I shall have my jolly spin in spite of them." Nor would you say: "I can take care of myself, and if I run into somebody else that is his look out." If you are an experienced cyclist you would keep on your seat, and go cautiously; if you are still a very inexperienced one, it would be wise to get off your cycle, and not mount again till you had come to the curve, and gone round it, and seen what is beyond.

The notice board is not an actual prohibition to go along the "King's highway" if you choose. The people who put up the board have no authority over you. But your own instincts of self-preservation, and I hope also your instinct of loyalty and good comradeship with the possible other cyclist who may be at the bottom of the hill, would suggest to you not to throw away the guardianship of a caution from those who know more than you do about the road.

Having given you this general indication of the principle which I am trying to explain, we will go back to the question of an imaginary working hypothesis.

My imagination, as I told you, showed me that my mind would travel quickly and easily along the road opened up by supposing that Israël means Rhythm.

Looking back in my memory, I could not find the smallest indication that anybody had either come to grief himself or offended any Hebrew person by behaving as if the people of Israël were the People of Rhythm; and there is nothing in the Ten Commandments to suggest that there is any harm in doing so. So I started off on a glorious, easy, rapid spin; and arrived, without any mishap, at several very interesting bits of scenery.

Now let us take the case of the old Scotch legend of Macbeth, as told by Shakespeare.

Macbeth and his wife appear to have been, at first, very well-intentioned, good people, as human beings go; better than most people; and enormously better than Jacob, or his mother, or his uncle, or most of the people belonging to him. Macbeth was a brilliant and successful soldier; his imagination suggested to him that he had it in him to rise rapidly to fortune and power. He might become Thane of Cawdor, and some day even King of Scotland. His imagination was so vivid that he pictured three old women going through some heathen incantation and predicting to him that he would be Thane of Cawdor and King. Here was a road open, along which it was quite sure that his mind would travel easily if he would let it do so. The question was: Should he let it go along that road? Now there were living at the time a Thane of Cawdor and a King of Scotland. While they lived, he could not be either. The commandments say, "Thou shalt not covet thy neighbour's goods." Here was a danger signal. If Macbeth had known as much as Shakespeare knew about the art of sound thinking, he would immediately have said to himself, " 'Cawdor' and 'King' are the roads that I had better not travel along just now, for fear the wheels of my mind should get too much way on, and carry me into danger." But Macbeth had either not learnt algebra at school, or, if he had, he had only crammed it up for examination out of a textbook, and not learned it as the Science of the *Laws of Thought.*

Another day his imagination showed him a dagger. A dagger is a thing to kill people with. As a soldier, he had probably used a real one in war. But, if he had had any proper nerve training, he would have known that when his imagination was so vivid that he did not, for the moment, know an imaginary dagger from a real one, he ought immediately to "go slack"; to lie down and think about the moors or the sky, or about anything or anybody that was not connected with doing anything in particular, with planning anything, with taking any resolution, and especially with breaking any of the Ten Commandments. He had already told his wife about the three old women. If she had been a sensible woman, she would have told him that she wanted to go away from home; and got him to take her right away for a few weeks; and kept him busy and amused in thinking of other things; till he left off seeing things that were not there. But neither Macbeth nor his wife knew as much as Shakespeare did about the value of danger signals and the conditions for making a safe working hypothesis.

You had better read the story of Macbeth and see for yourselves what they did do.

Next to the old Hebrew books, Shakespeare is the best road map that I know of for people who wish to travel safely about the country of the imagination.

Chapter 12

Jacob's Ladder

In Chapter X. I set you children a question:—Why did Jacob's angels come down a ladder, whereas other Hebrews saw angels mixed up with romantic pretty things such as wings and clouds?

I hope some of you have made a guess before now; but some are not good at guessing. I will tell you what may help you to find out.

If a bird wants to go up and down from the roof to the garden, it trusts to its wings. A man has to use a ladder: step,—step,—step.

If a bird is not fully fledged or has a broken wing, it has to find something more or less like a ladder; and go up and down bit by bit: hop,—hop,—hop.

If an artist wishes to draw a parabola, he does it freehand, that is to say, he just draws the curve He does not take all the trouble which Mrs Somervell's book makes little children take, of getting the curve step by step by the method of Finite Differences.

Jacob wished to be rich. Some angel, but a very bad one, inspired him with an idea of getting rich in one big sweep, by cheating his father and brother. By wanting to do things in that sort of quick, easy way, when he did not yet know how to do things both quickly and rightly, he got into terrible trouble and had to leave his country.

Now I suppose that the angels who converted him meant to say something like this: "It is all very well for good, holy, God-fearing men like your father and grandfather to go where they are taken by angels who can move about on wings; but you are at present a stupid, clumsy person; your wings have not grown yet, or you have broken them by being covetous. We are going to show you how *you* should go about: step,—step,—step. Have patience, and take pains; and don't go about on magic carpets."

Chapter 13

The Great x of the World

A great question which people like to quarrel about is:—Who or What made things be as they are? As soon as people grew clever enough to think about anything except scrambling for food and taking care of their own babies, they began quarrelling about Who or What made things be. Nobody knew anything about it; and most people had a great deal to say about it. Moses saw that there was no hope of getting a country orderly while all this confusion was going on; so he said to the Hebrews, "I must not allow all this confusion to go on among a people that I am made responsible for. None of us have ever seen the Maker of things. We can see the things growing, but not the force that makes them. *That* is our X; our Unknown. We are going to begin by stating that we don't know. We are going to call the Maker of things 'I Am,' or 'That which is, whatever it is'; and we are going to make two hypotheses to start with. We are going to try thinking of 'I Am' as Unity; one, and not several or a fraction. We will also try thinking of 'I Am' as No-Thing,—we are not going to suppose at present that any particular kind of thing made the rest; we will suppose that 'I Am' is not a thing. When we find that any particular proceeding or behaviour destroys men, or makes them too sickly or weak or stupid or quarrelsome to manage other creatures and keep the upper hand of the world, we will say, for short, that 'I Am' does not like or does not intend the people of Israël to go on with that kind of proceeding or behaviour.

"Now these two hypotheses are as much as we can deal with for the present. Anybody who wants to think out other hypotheses than those will have to think to himself, or go out of the country that I am to manage.

"Now we will arrange all the facts that we know round the statement of our own ignorance; and then try our hypotheses on them.

"We know that eating the flesh of certain uncleanly animals gives people certain diseases; we will say, for short, 'I Am' does not intend the Hebrews to eat the flesh of those animals. We know that if people are dirty in their habits and careless in preparing their food and in washing their hands before they touch food, they get fevers; we will suppose that 'I Am' does not intend the people of Israël to be dirty in their habits. We know that if people burn things

29

the smoke of which makes them drunk and silly, they manage their affairs badly, and make mistakes, and do not grow their crops properly, and are not ready to fight when enemies attack them. The people in neighbouring countries say that the Maker of things likes or dislikes to smell the smoke of these drugs; they know no more than we do what *He* likes to smell, but we are going to suppose that 'I Am' does not like *us* to smell them."

The Hebrews never found out what "I Am" is; but those who stuck loyally by the hypotheses of Moses, and refused to be distracted from the matter in hand, or to talk about anything except the experiment which they were trying, found out several things that were very useful to them. For instance, about weather and the electricity of the atmosphere, and how to take care of their health, and how to use their imagination to supply them with working hypotheses for a variety of sciences, and how to use their dreams to show them where they had been making mistakes and spoiling their brains. Whereas the people who would insist on shouting and arguing and quarrelling about things which were only wild guesses got on very slowly with learning Science.

Chapter 14

Go Out of My Class-Room

A story is told of one of the orderly pupils of Mosaism who got to know a good deal about weather and electricity; and at last he got out of patience with the people who wanted to shout and argue. And he said to them: "What is the good of all this arguing backwards and forwards about things that we do not know and cannot settle? Let us try a fair experiment. You go on shouting and doing whatever *you* think the Unseen Powers like; and I will do what *I* think will get them to do what *I* like. And let us agree that whichever of us can draw a spark out of a thundercloud shall be considered to know most about how to come to an understanding with 'I Am.'"

So the other people shouted and jumped about, and cut themselves with knives; because they had taken it into their heads to imagine that the Maker of things liked to see that kind of behaviour.

Why they thought so I cannot conceive. But there's no end to the rubbish that people get to think when they argue about what X is, instead of trying hypotheses in an orderly manner.

The Unknown Powers let them shout all day long; and then Elijah got a spark out of a thundercloud.

The same sort of thing happened again about a hundred and fifty years ago. Various sorts of priests were shouting and arguing about what "I Am" wished people to believe and to think; and then Benjamin Franklin and his friends, who had not been mixing up with the argument or making wild guesses, but quietly experimenting and dealing logically with the fact of their own ignorance, sent up a kite into a thundercloud, and got a spark down; and the consequence of that is that all kinds of people say, "What a wonderful man Benjamin Franklin was!" and all sorts of people are able to ride about in electric trams.

But the curious part of the matter is that many people use electric trams to go to meetings, on purpose to shout and argue and make wild guesses about things they know nothing about!

However, what they choose to do is not our business. You are living in an orderly school; and of course you do not argue about things you know nothing about. Let us go back to our Hebrew electrician.

31

He had shown the people of Israël what comes of sticking peaceably to one's working hypothesis. If he had been thoroughly logical he would have gone on sticking to it. He would have said to the people of Israël, "Now you see that I can teach you electricity; this land is going to be my class-room; make those shouting people hold their tongues, or else go away; so that we can go on with our lessons in peace. When they want to learn electricity properly, they can come back." But he was in too great a hurry to make a complete and final settlement. A good teacher sends a noisy, troublesome pupil out of his class-room for the time, but does not expel her from the school merely for being troublesome. The shouting people were among the facts which "I Am" put before Elijah to deal with. He found it necessary to eliminate them in order to reduce his data within the compass of his power to manage, but he should have done it as a provisional necessity. He should have arranged his mind on the understanding that the eliminated data would have to come back.

Instead of that he used his power and science to kill them; and gave way to the feeling that they were got rid of and done with.

And then his mind began to go wrong. He lost his nerve. He began to talk nonsense about things *he* knew nothing about, and led a great many people into mistakes.

Chapter 15

When you come to quadratic equations you will be confronted with an entity (or non-entity) whose name is written this way—$\sqrt{-1}$, and pronounced "square root of minus one." Many people let this nonentity persuade them to foolish courses. A story is told of a man at Cambridge who was expected to be Senior Wrangler; but he got thinking *about* the square root of minus one as if it were a reality, till he lost his sleep and dreamed that *he* was the square root of minus one and could not extract himself; and he became so ill that he could not go to his examination at all. Angels, and square roots of negative quantities, and the other things that have no existence in three dimensions, do not come to us to gossip about themselves; or the place they came from; or where they are going to; or where we are going to in the far future. They are messengers from the As-Yet-Unknown; and come to tell us where we are to go next; and the shortest road to get there; and where we ought not to go just at present. When square root of minus one comes to you, behave reasonably about him. Treat him logically, exactly as if he were six or nine; only always remember to keep well in front of you *the fact of your own ignorance.* You may never find out any more about him than you know now; but if you treat him sensibly he will tell you plenty of truths about your x's and y's, and other unknown things.

Please don't suppose that I have always behaved sensibly to angels. I have often made serious mistakes in dealing with them. I have acted in haste and have had plenty of reason to repent at leisure. But one thing they have taught me is that we need never be *afraid* of angels, whether white or black, as long as we keep the laws of logic. Another thing they have shown me is that angels never really gossip. They have often pretended to gossip to me; but I have found out afterwards that they have been talking clever nonsense in order to test me and prove me; so that I might see in what an illogical state of mind I have met them. Angels leave real gossip to old women who have done their life's work and have time to sit in the chimney corner and tell tales about their past experiences to their child friends.

Chapter 16

Infinity

You remember the angel who looks like this, $\sqrt{-1}$. Now I am going to introduce you to another angel. It is called "Infinity." When you come to it, remember what I told you before—Angels are messengers from the great world of the "As-Yet-Unknown." They never gossip about their private affairs, or those of other angels. They come to tell you either about what you are to do next, or about something you had better not do next; and if you ask them impertinent questions about things that do not concern you for the time being, they will give you headaches and make your head spin: just to teach you to mind your own business. This particular angel always comes with a message about a broken link or a loosened chain. It comes, when an hypothesis has been fully worked out, to tell you that you are now free from the bonds of that hypothesis and at liberty to start experimenting on a fresh one. But its message is never: "You have got out of that particular house of bondage and therefore you may, for all the rest of your life, run riot, and eat, and drink, and do, whatever you please." Its message always is: "You have outgrown that master; now you may take a holiday and have a fling before you go into a higher class; but, just because you are set free, look out for danger traps; and mind your Ten Commandments."

You will understand Infinity's messages better if you will read carefully what is written about it in Chapter XV. of "The Logic of Arithmetic." It brought the answer to the question: "How many children could pass through a school-room without the apples all being eaten up, supposing that none of the children ate any?"

Let us go over that ground again. Suppose there is a cake on the table. How many children can go through the room without the cake being all eaten up?

Well, that depends on two things: the size of the cake, and the share which each child eats. If the cake weighs two pounds, and each child eats two ounces, it will be all eaten up when sixteen children have gone through the room. If the cake weighs only one pound, it will be eaten up when eight children have gone through the room. But if each child eats only one ounce, then again sixteen children will have to go through the room before the cake is eaten up, and so on. Many questions could be asked, all depending on the size of the cake and the size of each child's share.

All this time you are tied to an hypothesis that the children eat cake (more or less).

But now suppose we are freed from that hypothesis. Suppose no cake is given to the children. How many can pass through the room before it is all eaten up?

The answer to that is: "An infinite number." Infinity does not mean any particular number, or a very large number. It means a loosened chain, a discarded hypothesis, escape from the rule we were working under. Something else, not the size of the cake, determines the number of children. Infinity does not mean that there are enough children in the world now to go on passing through the room *for ever*, but that the number of children who pass through the room, now that the share of each child is 0 (zero), will have to be determined by the number of children that there are in the school, or the parish, or wherever it is that the children are supposed to come from; *and not by the size of the cake*. The size of the cake has no longer anything to do with answering the question: "How many children can pass through the room before the cake is all eaten?"

Chapter 17

From Bondage to Freedom

Moses had said, from the first, that the people of Israël would have to think of "I Am" as the deliverer from bondage; but they were not, at the time when he said it, advanced enough in their algebra to understand that idea properly. So he gave them, as an hypothesis to work on for the time being, that "I Am" did not like the *people of Israël* to eat and drink and smell unwholesome things. He wished to make them attend to their own affairs, and think as little as possible about what was done and thought outside of their own land.

But, after the time of Elijah, there came a change. A higher kind of algebra came into use. Its incarnation was called Isaiah.

The imagination of the Hebrews broke loose from the hypothesis that "I Am" had wishes and likes about the people of Israël different from what was right for all the rest of the world.

When that hypothesis was taken away, the imagination of such people as Isaiah took wings and flew to—well—we do not know where, but we call it Infinity. We know nothing about Infinity; except that it comes when a chain is loosened.

If you want to understand what it was that happened to Isaiah, and what Infinity means in algebra, this is how you can find out. Get a bowl and dip up some of the water out of a barrel in which a gnat has laid her eggs. Little wigglers are born from those eggs. If you watch them you will see that they swim in different positions, some with their tails uppermost, some with their heads uppermost. There may also be some worms, who do not swim much, but wriggle about at the bottom of the bowl. Perhaps if we could hear them talking we should hear them quarrelling about which was the right position. Some of them might be disputing about what would happen to them in the future. They might quarrel till the end of the world, and know no more about it at the end than at the beginning. They are all tied by the same hypothesis:—that everybody lives under water. It is a very good working hypothesis for them; for if one of them got out of the water it would die. If they knew algebra properly, they would understand that water is only their present working hypothesis, and that it is quite possible there may be people who live out of it. But it is not sure that they do know enough algebra to be *aware of their own ignorance.*

If you watch them carefully, you will some day see a wiggler come out of the water. He has got wings. The water-hypothesis no longer concerns him. Some link in the chain that bound him down to water has opened; he is set free; Infinity has come to him.

That is what happened to Isaiah when he got out of the kind of Mosaism by which such people as Joshua and Samuel were tied down. That is what will happen to you (if you learn your algebra properly) when you are no longer tied down to a, b, c, and $\sqrt{-1}$, as the values of x; and learn to see that the answer to a problem may sometimes be

$$X = Infinity.$$

Please notice that if a winged gnat fell back into the water he would die. You will find this a good working rule:—Whenever anything comes near your imagination which calls itself either "Infinity" or "The Liberation from Bondage," go slack for a few minutes; say over the Ten Commandments; and make a mind-picture of the gnat-grub in the water. Tell yourself that his best chance of growing strong wings and being able to fly, when Infinity comes and calls him to go up higher, is to stay in the water till the wings have grown strong and work out the water-hypothesis to its logical conclusion.

Then make another mind-picture:—The gnat who has got wings, and *therefore must not try to amuse himself in the water.*

Please observe:—There is nothing in this rule contrary to any commandment. Moreover, there is nothing slavish or degrading in it; nothing in the least like giving up your own liberty, or hampering your own initiative, or being a slave to past ages; nothing which prevents your being up to date and fit for the generation to which you belong.

You are not asked to have any opinions about it; or to think that it is a duty in itself; or to think that you are better than other people because you do it, or that every one is wrong who does not do it. If you do it, it will be for no reason that you know of, except that an old woman who has been trying to amuse you asks you to do it as a token of friendly feeling towards her.

Chapter 18

Appendix

The essential element of Algebra:—the habitual registration of the exact limits of one's knowledge, the incessant calling into consciousness of the fact of one's own ignorance, is the element which Boole's would-be interpreters have left out of his method. It is also the element which modern Theosophy omits in its interpretation of ancient Oriental Mind Science.

Men who wish to exploit other men fear nothing in logic or science except this element. They fear nothing in earth, heaven, or hell, so much as a public accustomed to realise exactly *how much has been proved, and where its own ignorance begins.* Exploiters fear this about equally, whether they call themselves priests, schoolmasters, college dons, political leaders, or organisers of syndicates and trusts.

As long as general readers can be kept from the habit of registering at every step the fact of their own ignorance and the limits of their own knowledge, a clever charlatan can deceive them about anything he pleases:—"from pitch-and-toss to manslaughter"; from Zero to Infinity; from the contents of a meat tin to the contents of an engineer's report; from the interpretation of a bill before Parliament to the interpretation of Isaiah.

Once get any fair proportion of the public into the steady habit of algebraising ignorance, and you will have done much towards reducing all kinds of parasitic creatures to the alternative of starvation, suicide, or earning their own living by rendering some kind of real service to the organism which supports them.

Logic Taught by Love

RHYTHM IN NATURE AND IN EDUCATION

A set of articles chiefly on the light thrown on each other
by Jewish Ritual and Modern Science

By

Mary Everest Boole

Crown 8vo, Cloth, 3s. 6d. net.

LIST OF CHAPTERS

1. In the Beginning was the Logos.
2. The Natural Symbols of Pulsation.
3. Geometric Symbols of Progress by Pulsation.
4. The Sabbath of Renewal.
5. The Recovery of a Lost Instrument.
6. Babbage on Miracle.
7. Gratry on Logic.
8. Gratry on Study.
9. Boole and the Laws of Thought.
10. Singular Solutions.
11. Algebraizers.
12. Degenerations towards Lunacy and Crime.
13. The Redemption of Evil.
14. The Science of Prophecy.
15. Why the Prophet should be Lonely.
16. Reform, False and True.
17. Critique and Criticasters.
18. The Sabbath of Freedom.
19. The Art of Education.
20. Trinity Myths.
21. Study of Antagonistic Thinkers.
22. Our Relation to the Sacred Tribe.
23. Progress, False and True.
24. The Messianic Kingdom.
25. An Aryan Seeress to a Hebrew Prophet.
    Appendix I.
    Appendix II.

London: C.W. DANIEL, 11 Cursitor Street, E.C.

Printed in the United Kingdom by
Lightning Source UK Ltd., Milton Keynes
137674UK00001B/325/P

9 781603 861267